Measure Up!

by Margie Burton, Cathy French, and Tammy Jones

Table of Contents

What Can You Measure? . . . . page 2

What Do You Use to Measure How
Long or How Tall Something Is? . . page 4

What Do You Use to Measure
How Much Something Weighs? . . . page 7

What Do You Use to Measure How
Hot or How Cold Something Is? . . page 10

How Can Measuring Help You? . . page 15

What Can You Measure?

You can measure how long or how tall something is. You can measure how much something weighs. You can measure how hot or how cold something is. These are some things that you can measure.

What Do You Use to Measure How Long or How Tall Something Is?

You can use a ruler or a tape measure at school to measure how long something is. You can use a ruler or a tape measure to measure how tall something is, too.

I am taller than she is.

Sometimes, you use paper clips
to measure how long something is.

Sometimes, you use your foot
to measure how long something is.

You can use these cubes
to measure how long something is, too.

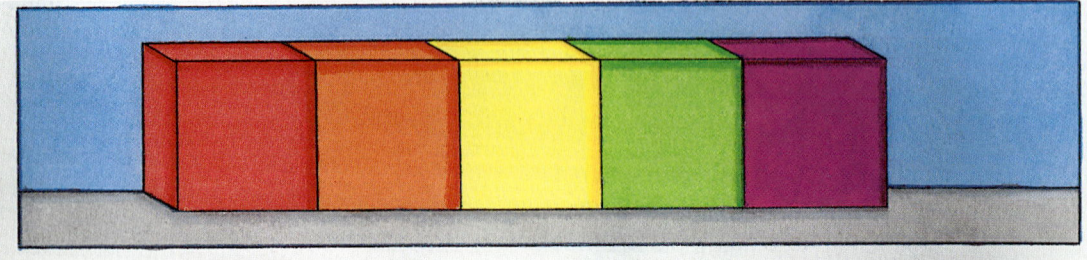

What Do You Use to Measure How Much Something Weighs?

You can use a scale to measure how much something weighs.

My dog weighs 25 pounds!

You can use a scale at home. You can use a scale at the store, too. You can weigh the vegetables at the store.

The vegetables weigh nine pounds.

This man uses a scale at work.
He weighs the box before
he puts it on the truck.

What Do You Use to Measure How Hot or How Cold Something Is?

You can measure how hot or how cold it is outside with a thermometer.

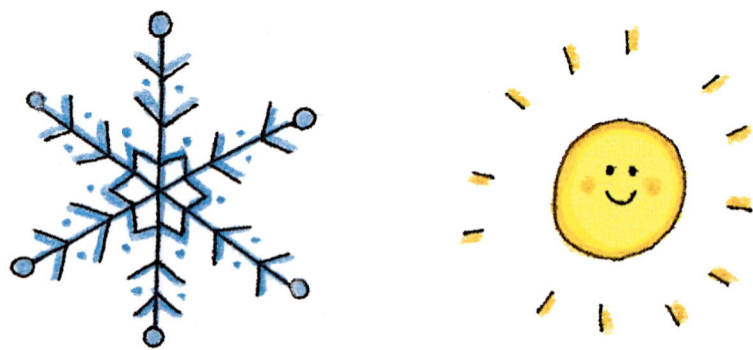

You know what to wear when you look at the thermometer.

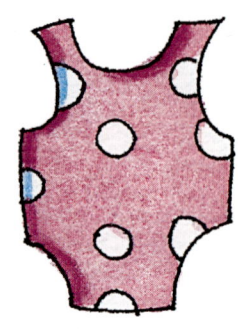

I look outside at the thermometer to know what clothes to wear.

You can tell the season by keeping track of the temperature over time.

My dad cooks with a thermometer.
He measures how hot or how cold
the food is.

My mom uses the thermometer, too.
It can measure how hot my body is.
If it is too hot, she knows
that I am sick.

How Can Measuring Help You?

You measure your foot to know what size shoes to buy.

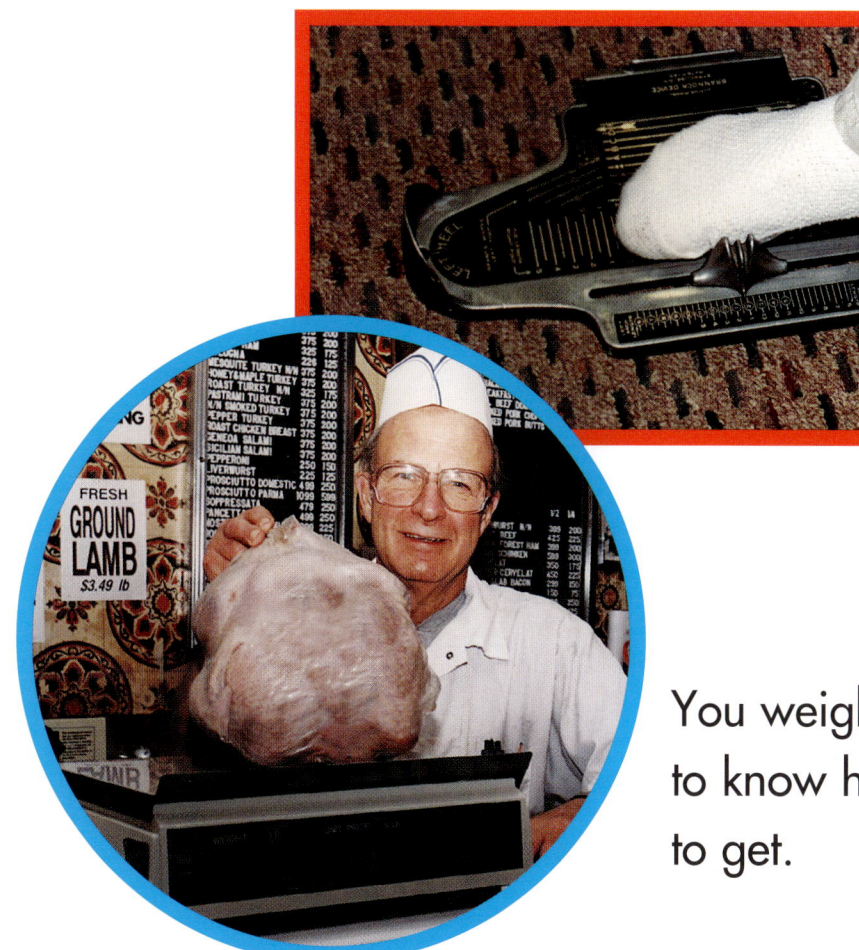

You weigh the meat to know how much to get.

I measure, too.

I measure the salt to bake some bread.